SNAILS

Rebecca Woodbury, Ph.D., M.Ed.

Gravitas Publications Inc.

SNAILS

Illustrations: Janet Moneymaker

Snails
ISBN 978-1-950415-61-8

Published by Gravitas Publications Inc.
Imprint: Real Science-4-Kids
www.gravitaspublications.com
www.realscience4kids.com

RS4K

Photo credits: Cover & Title Pg: Krzysztof Niewolny on Unsplash; Above, Alexas_Fotos from Pixabay; P.3. AB-7272, AdobeStock; P.5. Savo Ilic, AdobeStock; P.9. Krzysztof Niewolny on Unsplash; Waugsberg, CC BY SA 3.0; P.11. Alexas_Fotos from Pixabay; P. 13. Left, Illustrations by Debivort at en.wikiped, CC BY SA 3.0; Right, Robert Hershler & Hsiu-Ping Liu, CC BY SA 3.0; P.15. Gabriela Fink from Pixabay; P.19. Top, Jess Van Dyke, Snail Busters, LLC, Bugwood.org; Left, Mandy Lindeberg, NOAA/NMFS/AKFSC; Right, István Mihály from Pixabay

If you have a garden, you might find that some of the plants have holes in their leaves.

I wonder what made those holes?

It might be that a **snail** has
been snacking in your garden.

Hello
down there.

Hello,
snail!

Snails are a type of **mollusk.**

Review: MOLLUSK

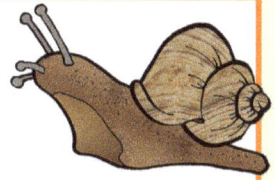

Mollusks are soft-bodied animals that live in oceans and lakes and on land.

All mollusks are in the group called **Mollusca.**

There are three different types of mollusks in the group Mollusca.

Gastropods ------→ Snails and slugs

Bivalves ------→ Clams and oysters

Cephalopods --→ Octopuses and squids

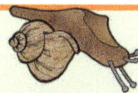

Review: MOLLUSK

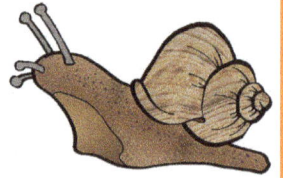

- All mollusks have **organs.** An organ is a body part that has a particular job to do in the body, like the heart, stomach, and brain.

- Mollusks have a **mantle** that covers the organs, and they have one or more feet or arms. Most have a shell either on the inside or outside of the body.

A snail has a **foot,** which is
a large muscle used for moving.

Foot

A snail has eyes at the tip
of stalks called **eyestalks.**

I wonder what it's
like to wave your
eyes around?

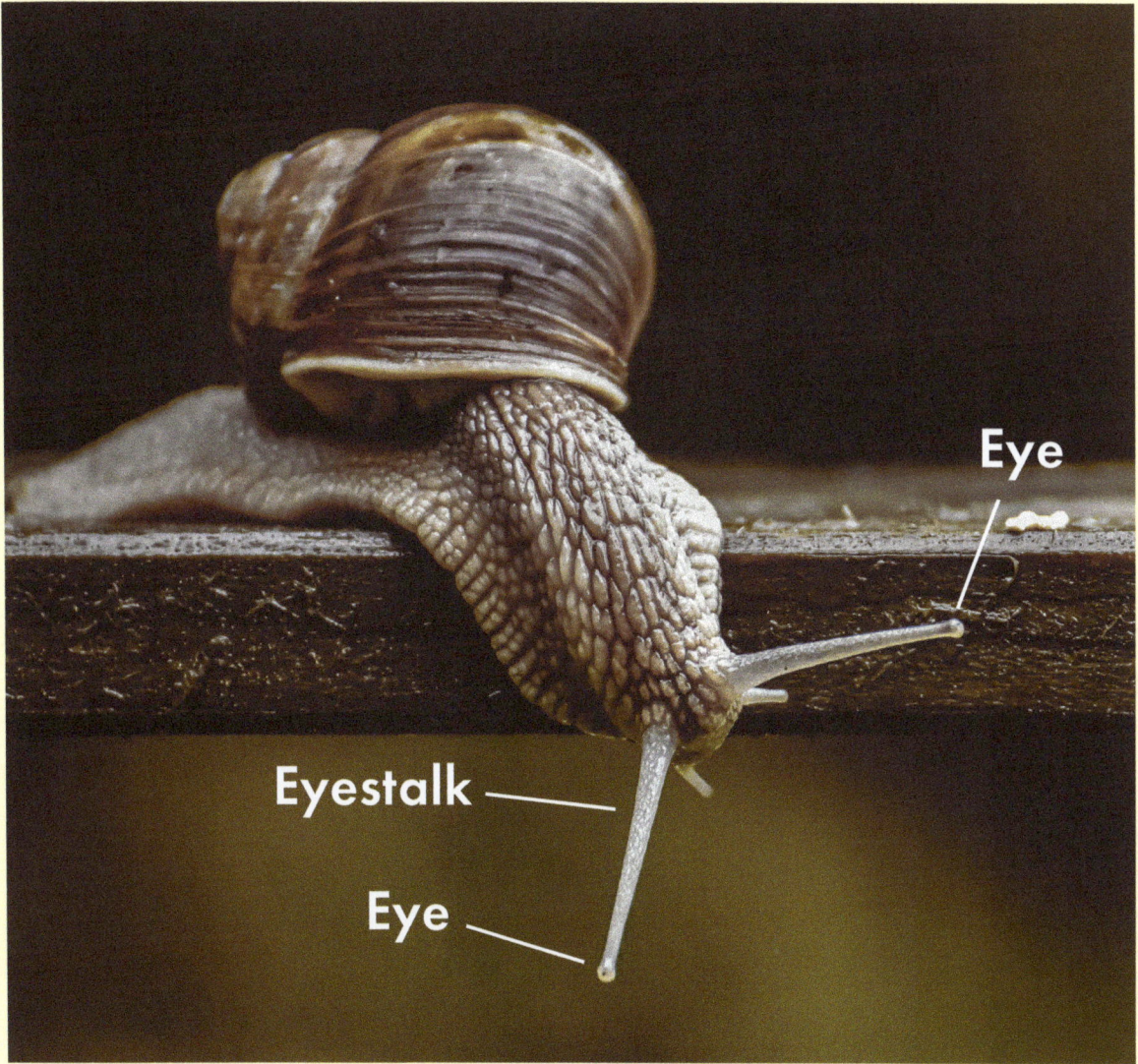

Eye

Eyestalk ———

Eye ———

A snail has a toothed tongue called a **radula** that is used for eating food.

I wonder why a snail doesn't cut itself with its tongue?

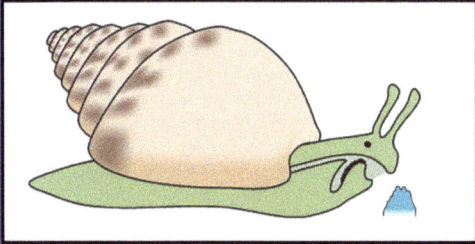

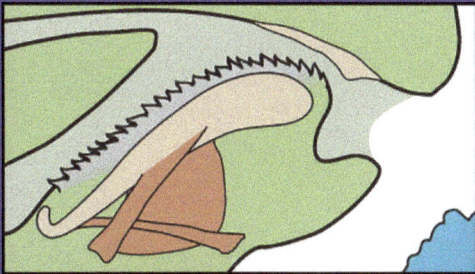

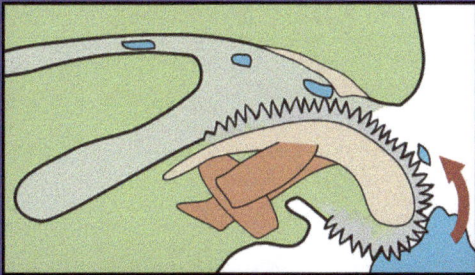

A snail uses
its radula to
scrape up food

Teeth on a radula
seen with a
microscope

A snail carries a spiral
shell. The snail can pull
itself into its shell for safety.

I do not think I
would like to carry
my house with me
everywhere.

Nope.

Snails move slowly and leave a trail of slime behind them. The slime helps a snail move along a surface while protecting the snail's body from sharp objects. It also helps the snail stick to objects as it climbs them.

The slime looks sparkly in the sunshine.

There are many different kinds of snails. Some live on land and some live in water.

Some snails live in lakes and ponds.

Some snails live in oceans.

Some live on land.

In some parts of the world
people eat snails and
think they are yummy!

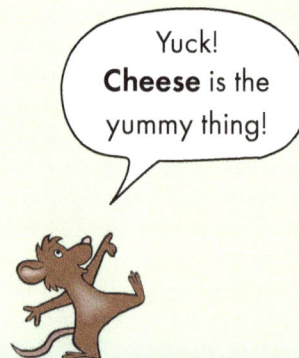

Yuck! **Cheese** is the yummy thing!

How to say science words

bivalve   (BIY-vaalv)

brain   (BRAYN)

cephalopod   (SEH-fuh-luh-pahd)

clam   (KLAAM)

eyestalk   (IY-stawk)

foot   (FUHT)

gastropod   (GAA-struh-pahd)

heart   (HAHRT)

mantle   (MAAN-tuhl)

microscope   (MIY-kruh-skohp)

Mollusca   (mah-LUH-skuh)

mollusk   (MAH-luhsk)

muscle   (MUH-suhl)

octopus   (AHK-tuh-puhs)

organ   (AWR-guhn)

oyster   (OY-stuhr)

radula   (RAA-juh-luh)

science   (SIY-uhns)

slug   (SLUHG)

snail   (SNAYL)

spiral   (SPIY-ruhl)

squid   (SKWID)

stomach   (STUH-muhk)

tongue   (TUHNG)

www.ingramcontent.com/pod-product-compliance
Lightning Source LLC
Chambersburg PA
CBHW040152200326
41520CB00028B/7578